WHEN I GO TO THE BEACH

Miranda Kelly

Crabtree Publishing
crabtreebooks.com

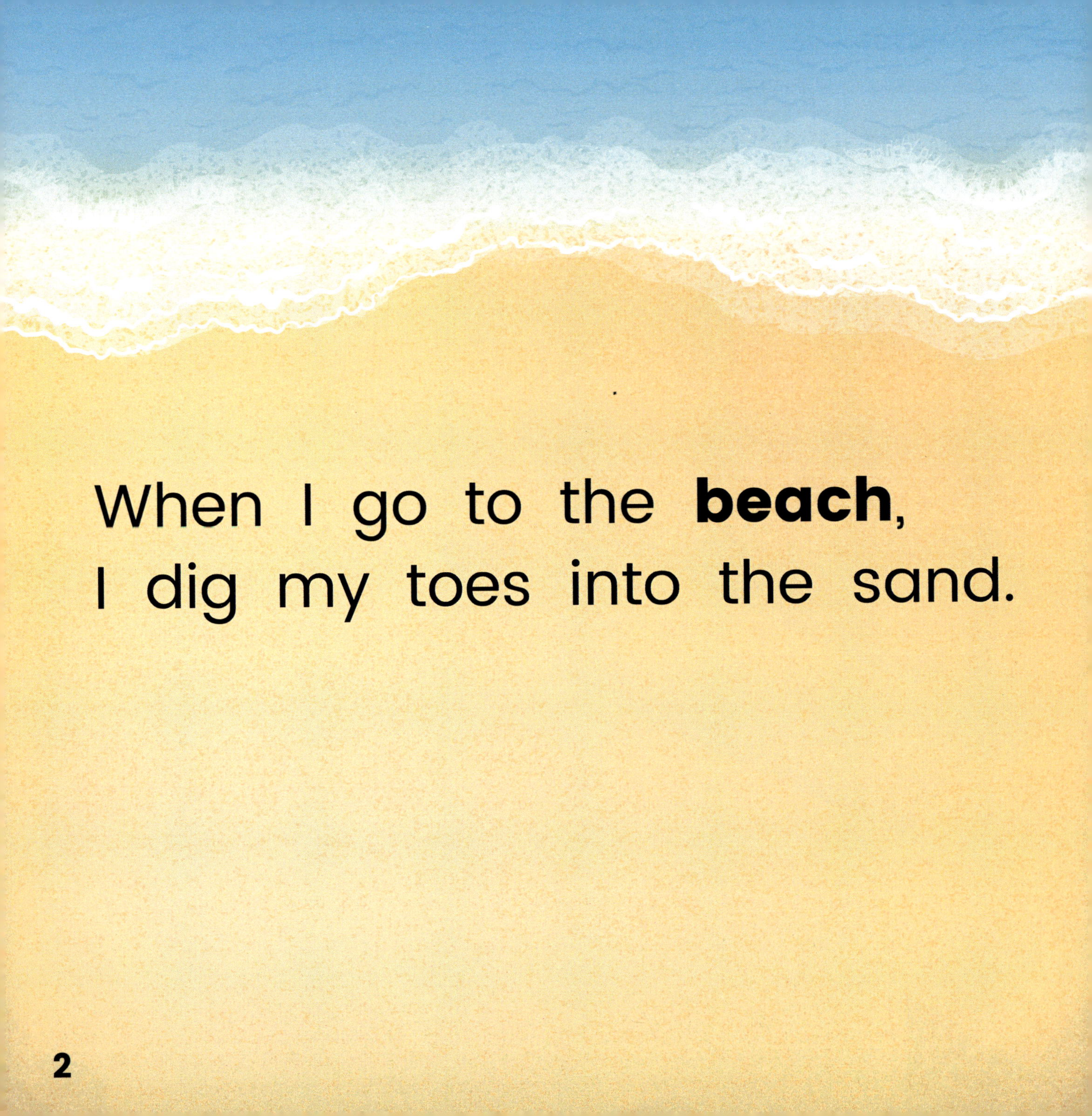

When I go to the **beach**,
I dig my toes into the sand.

I **build** sand castles.

I **listen** to the waves.

I splash in the water.

I **collect** shells.

I look for **crabs**.

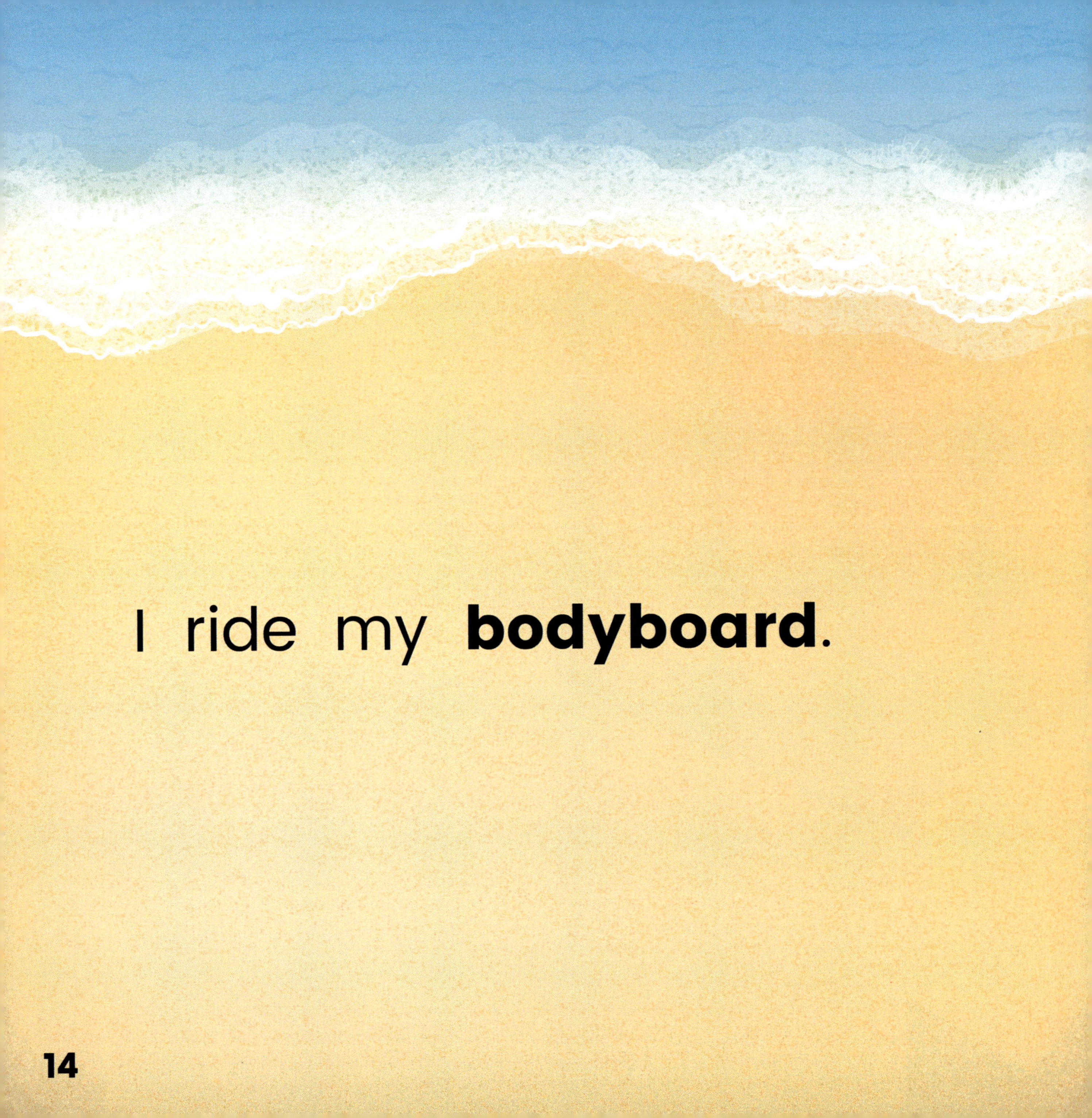

I ride my **bodyboard**.

I meet with my friends.

I watch the sun set.

I have fun when I go to the beach.

Glossary

beach (BEECH): A beach is an area of sand or pebbles that lies along the edge of a lake or the ocean.

bodyboard (BOD-ee bord): A bodyboard is a short, foam board that people can lie on to ride waves.

build (BILD): When you build something, you put different parts together.

collect (kuh-LEKT): When you collect things, you gather them.

crabs (CRABS): Crabs are animals with eight legs, two pincers, and a hard shell. There are many different kinds of crabs.

listen (LISS-uhn): When you listen, you pay attention so you can hear something.

Index

crabs 12
sand 2, 4
sand castles 4
shells 10
waves 6

School-to-Home Support for Caregivers and Teachers

Crabtree Seedlings books help children grow by letting them practice reading. Here are a few guiding questions to help the reader with building his or her comprehension skills. Possible answers are included.

Before Reading

- What do I think this book is about? I think this book is about having fun at the beach.
- What do I want to learn about this topic? I want to learn about the different animals people might see at the beach.

During Reading

- I wonder why... I wonder why the boy listens to the waves. What do they sound like?
- What have I learned so far? I have learned that people might see crabs at the beach.

After Reading

- What details did I learn about this topic? I learned that going to the beach is fun. Children can build sand castles, collect shells, and play in the water.
- Read the book again and look for the vocabulary words. I see the word ***build*** on page 4 and the word ***collect*** on page 10. The other vocabulary words are found on pages 22 and 23.

Crabtree Publishing

crabtreebooks.com 800-387-7650

Print book version produced jointly with Blue Door Education in 2022

Content produced and published by Blue Door Publishing LLC dba Blue Door Education, Melbourne Beach FL USA.

Written by Miranda Kelly
Production manager: Candice Campbell

Hardcover 978-1-4271-2960-4
Paperback 978-1-4271-2970-3

Printed in the U.S.A./CP082026

Library and Archives Canada Cataloguing in Publication
Title: When I go to the beach / by Miranda Kelly.
Names: Kelly, Miranda, 1990- author.
Description: Series statement: In my community | "A Crabtree seedlings book". | Includes index.
Identifiers: Canadiana 20200388177 | ISBN 9781427129604 (hardcover) | ISBN 9781427129703 (softcover)
Subjects: LCSH: Beaches—Juvenile literature.
Classification: LCC GB453 .K45 2021 | DDC j551.45/7—dc23

Published in Canada
Crabtree Publishing
616 Welland Avenue
St. Catharines, Ontario
L2M 5V6

Published in the United States
Crabtree Publishing
347 Fifth Avenue
Suite 1402-145
New York, NY 10016

Photo credits: Cover: Beach type © dgbomb, cover photo © NadyaEugene; cover and page background © art_of_sun; page 3 © Natee K Jindakum; page 5 © Darren Green; page 7 © Christin Lola; page 9 © Pavel_Bogdanov; page 11 © mirtmirt; page 13 © kavida; page 15 © NadyaEugene; page 17 © Tropical studio; page 18 beach © Galyna Andrushko, page 18 bodyboard © Evgeniy pavlovski; page 18 sand castle © Alex Tor; page 19 crab © jiraphoto All images from Shutterstock.com; page 21© istock.com/Rawpixel; page 23 ©istock.com/Mimadeo

Library of Congress Cataloging-in-Publication Data
Names: Kelly, Miranda, 1990- author.
Title: When I go to the beach / by Miranda Kelly.
Description: New York, NY : Crabtree Publishing, 2021. | Series: In my community, a Crabtree seedlings book | Includes index.
Identifiers: LCCN 2020050814 | ISBN 9781427129604 (hardcover) | ISBN 9781427129703 (paperback)
Subjects: LCSH: Beaches--Juvenile literature.
Classification: LCC GB453 .K45 2021 | DDC 428.1--dc23
LC record available at https://lccn.loc.gov/2020050814